60
ROUGH
ROAD
NEXT 2 MI.
ROUTE
66

ROUTE US 66

Publications International, Ltd.

Images from Shutterstock.com

Louis Weber, CEO
Publications International, Ltd.
8140 Lehigh Avenue
Morton Grove, IL 60053

ISBN: 978-1-64558-257-1

Manufactured in China.

8 7 6 5 4 3 2 1

Let's get social!

 @Publications_International

 @PublicationsInternational

www.pilbooks.com

Table of Contents

The Mother of All Road Trips

Route 66 is not remembered simply as a conduit between Chicago and Los Angeles, but as a passage to the heart of American cool—the Classic Road Trip with a side helping of rock 'n' roll. With the convertible top down and the sun-soaked beaches of paradise somewhere over the horizon, day-to-day life was temporarily forgotten in the rearview mirror. Because the road snaked right through the heart of numerous cities and towns, the opportunities for fun (and trouble) were plentiful.

Now represented by a patchwork of interstates, two-laners, and decrepit stretches of asphalt, old Route 66 doesn't really exist, at least in the eyes of Congress or mapmakers. But it still exists for the nostalgic tourists—an unusual mix of bikers, travelers, and RVers—who stick as close as possible to the roads that were once Route 66 and see the old sights that have withstood the test of time.

The appearance of Route 66 came as average Americans were gaining access to personal automobiles. It symbolized freedom, mobility, and optimism, and became the emblem of a new type of American vacation—the road trip.

CHICAGO
ST.LOUIS
HOLBROOK
SANTA FE
OKLAHOMA CITY
SPRINGFIELD
L. A.
ALBUQUERQUE
AMARILLO
TULSA
ROUTE
66
JOLIET
ON
66

The Mother Road, in Miles

The total length of Route 66 varied over the years as bypasses and improvements altered the tally. Complicating this is the fact that there was more than one version of the highway at the same time in some places. The original route of the commissioned highway, as of 1926, is commonly cited as being 2,448 miles long. By 1947, alterations had brought the total mileage down to under 2,300.

- The Mother Road
- The Main Street of America
- The Great Diagonal Way
- The Road of Flight
- The Will Rogers Highway

State	1926	1947
Illinois	301 mi	284 mi
Missouri	317 mi	316 mi
Kansas	13 mi	13 mi
Oklahoma	432 mi	404 mi
Texas	186 mi	175 mi
New Mexico	487 mi	379 mi
Arizona	401 mi	395 mi
California	314 mi	314 mi

CHAPTER 1
Illinois

Just one street among many within the dense traffic grid of metropolitan Chicago, Route 66 begins as little more than a nostalgic idea. The route originally ran west through downtown, then turned southwest on Ogden Avenue, angling out of the city and through the suburbs. Cicero, Lyons, McCook, Burr Ridge, Bolingbrook . . . the sprawling suburbs and outlying communities now seem to go on and on. Even in the 1930s, leaving the urban landscape behind took time.

Eventually the road opened up and travelers entered a classic Midwestern landscape. Much of central Illinois is a pancake-flat vista of rich agricultural land. Motorists would have driven through mile after lulling mile of waving cornfields dotted with rural villages. Dwight, Pontiac, Lexington, and other farm towns were mere breaks in the endless corn.

At well over the 200-mile mark, Route 66 arrived in the Illinois state capital of Springfield, home of the Abraham Lincoln Presidential Library and Museum (and other Lincoln-oriented attractions). It traveled south through town, but soon the urban respite was over and the highway was swallowed again by the vast fields of the Corn Belt. The occasional town appeared—Thayer, Girard, Staunton, Hamel—but blink and you've missed them!

Downstate, the route angled southwest toward the Mississippi River. The route's entrance into Missouri changed over time, but all routes crossed bridges over the river into or close to St. Louis.

Chicago's tallest building, Willis Tower, looms behind the sign.

Downtown Chicago's Buckingham Fountain is just a short walk away from the eastern terminus of Route 66. It was completed a year after the official designation of Route 66.

Growing Pains

In the popular imagination, Route 66 was physically fixed and unmoving. But the historian of the Mother Road knows that there were quite a few variations in the actual route over time. As vehicles, economics, and transportation logistics changed over time, so did the route. The story of the Illinois section of Route 66 is one of evolution. With this in mind, it's worth noting that the spots we visit in this book may be from any of the route's historical alignments, or a short distance away.

1926. The highway connecting Chicago and St. Louis that preceded Route 66 was known as State Bond Issue 4. This patchwork of existing roads became the basis for the 1926 alignment. It was more winding than later routes, and frequently passed directly through many small towns. Road width was commonly 18-20 feet.

The 1930s. With car speeds routinely approaching 70 m.p.h., municipalities found the need to impose speed limits. Thanks to heavy traffic, the route was modified to bypass large, congested urban sections like Springfield and Joliet. Rest areas began springing up along the route.

WWII and beyond. The significant weight of heavy wartime traffic wreaked havoc on the route's paved surface. The Federal Defense Highway Act of 1941 provided cash for much-needed upgrades. As the route was repaired, lanes were widened and road engineering was improved, allowing for a safer drive.

A stretch of old Route 66 outside Pontiac—abandoned but still drivable.

(Top left) **A typical (unofficial) starting point of the route is at the corner of Adams Street and Michigan Avenue, across the street from the Art Institute of Chicago.**

(Top right) **The route passes below the shadow of Willis Tower.**

(Left) **The route runs by Chicago's venerable Union Station, a major rail transportation hub that has been in existence since 1925.**

(Right) The Midewin National Tallgrass Prairie, just south of Joliet, consists of over 18,000 acres of native tallgrass prairie ecosystem in the process of being restored.

(Bottom) The old prison in Joliet, famously featured in *The Blues Brothers*, closed in 2002. You can still get in though—the facility began conducting tours in 2018.

(Top and bottom left) **The Rich and Creamy ice cream shop is located on an old leg of Route 66 in Joliet. Blues Brothers Jake and Elwood hold down the roof.**

(Top right) **The Gemini Giant is a landmark statue adjacent to Route 66 just outside Wilmington.**

The Polk-a-Dot Drive In in Braidwood has been serving up classic diner food like burgers, fries, and shakes since 1962.

In Gardner, stop off for a moment of incarceration at the Two-Cell Jail. The tiny holding pen was built in 1906 and is nothing more than a single room divided by bars. Visitors are encouraged to stop and lock themselves up for a quick photo.

(Top left) **Ambler's Texaco Gas Station sits in restored 1930s-40s glory in the Village of Dwight. The station's quaint, cottage-like design is typical of many early midwestern gas stations.**

(Top right) **Dwight's Oughton Estate Windmill has been standing above the town since 1896.**

(Left) **The Standard Oil Gas Station in Odell was built in 1932. It operated until the 1970s.**

(Right) A favorite Route 66 mural for the Illinois section is located in downtown Pontiac.

(Bottom) Towanda is the site of the famous Dead Man's Curve. Drivers unfamiliar with the route often misjudged the sharp turn. According to locals, motorists making the trip from Chicago seemed particularly susceptible to accidents.

The silhouette of a motorcycle cop from days gone by stands vigilantly outside Pontiac. The Illinois State Police ended motorcycle patrols around 1949.

The Illinois State Police came into existence in 1922. That year, Len Small was elected state governor. He campaigned on the slogan "take Illinois out of the mud," and the coming years would see significant improvements to the state's road infrastructure.

The first state troopers wore surplus World War I gear and drove motorcycles. At first, they focused on truck regulation and tried to enforce weight limits. By 1927, some officers were driving patrol cars. Cars were upgraded with radios in the mid-1930s.

Traffic on Route 66 increased steadily, and with that traffic came new challenges: speed-seeking citizens. Deaths from highway accidents were something new in the state and the police began focusing on reducing the problem. Crashes became so frequent that the highway acquired the nickname, "Bloody 66." Though the moniker has faded from public memory, it illustrates the fact that the country had to learn how to create safe high-speed highways and motorists had to learn to abide by traffic laws. One officer noted that most accidents occurred where the route intersected with county roads and railroad crossings. In Illinois, two stretches of highway came to be called "Dead Man's Curve" and "Dead Man's Alley." But it wasn't just the Illinois portion of Route 66 that was dangerous. Arizona's slice of the route may have been the worst of all. And in 1941, over 450 people died along a deadly section in central Missouri. Even New Mexico's *Gallup Independent* called Route 66 "the state's most dangerous highway" in 1953.

(Left) A 19-foot-tall Paul Bunyon (the misspelling is deliberate) stands proudly in downtown Atlanta. Originally hailing from a hotdog stand in Cicero, the statue was moved to its new location in 2003.

(Top) The tiny town of McLean is home to the Dixie Truckers Plaza. The restaurant opened in 1928 to sell sandwiches to truckers. Little more than a 6-stool kitchen counter when it opened, it became a major stop with cabins and a cattle pen by the 1930s. The restaurant is still open today.

Known as the geographic center of Route 66 in Illinois, the town of Atlanta features a quaint and classic downtown area with a Norman Rockwell vibe.

The Abraham Lincoln Family Home in Springfield is where Lincoln lived (from 1844 to 1861) before becoming president.

(Top) **The Lincoln Tomb and Monument in Oak Ridge Cemetery.**

(Right) **A statue of Abraham Lincoln stands in front of the state capitol building in Springfield. The capitol features a number of Lincoln attractions, including the Abraham Lincoln Presidential Museum, Lincoln Tomb and Monument, and Lincoln Home National Historic Site.**

(Top left) A unique kind of road surface appears in Auburn: over a mile of hand-laid brick.

(Top right) Tiny Carlinville's Marvel Theatre was first opened in 1920.

(Left) Litchfield's family-run Ariston Café has been serving up diner favorites like steaks, BLTs, and bacon and eggs for decades.

(Top left) Henry Soulsby used his life savings to build this service station in Mount Olive in 1926, just in time for the birth of Route 66. Preservation efforts began in 2003, and the station now appears as it would have after World War II, when it was at its busiest.

(Top right) The McKinley Bridge was built in 1910. It was Route 66's original crossing point into Missouri.

(Right) The Cahokia Mounds State Historic Site lies between Collinsville and East St. Louis. The remains of this ancient Native American city cover over 2,000 acres.

The old Chain of Rocks Bridge became part of Route 66 in 1936. This memorable bridge features a 30-degree change of direction in the middle. It first opened to traffic in 1929.

(Right) **The Chain of Rocks Bridge is now a picturesque and automobile-free crossing for bikers and hikers.**

(Bottom) **Across the Mississippi River, St. Louis, Missouri, beckons.**

CHAPTER
2
Missouri

After the sleepy farm towns of Illinois, the jaunt over the Mississippi River into bustling St. Louis is a bracing change of scenery. A northern leg of Route 66 across the Chain of Rocks Bridge bypassed the urban core, but other routes went through or near downtown. There are so many route options in downtown St. Louis that the best may simply be the one that pleases you the most.

Heading southwest out of the city, route travelers are presented with something new: hills! Missouri's roughly 300-mile route offers plenty of rolling woodlands and lush countryside. The Ozark Highlands in the south of the state provide the only significant hills on the route east of the Sandia Mountains in New Mexico.

The Gateway Arch was completed in 1965 and is the tallest arch in the world. The arch stands 630 feet tall and 630 feet wide. It takes about 4 minutes to ride a tram to the top. The observation area at the top of the arch features 16 windows, giving views to the east and west.

St. Louis Side Trips
Anheuser-Busch Brewery
Busch Stadium
Crown Candy Kitchen
Gateway Arch
Missouri Botanical Garden
Missouri History Museum
National Museum of Transportation
St. Louis Art Museum
St. Louis Zoo
Ted Drewes Frozen Custard
Union Station

Stop Off in St. Louis

As one of the biggest cities along the route between Chicago and Los Angeles, St. Louis is worth slowing down for. It has always been a center of commerce and trade as well as a transportation hub. Primely positioned near where the Illinois and Missouri rivers join the Mississippi, it began existence as a trading place on the continent's central water highway. With the advent of the steamboat era, the city bloomed. Next came the transcontinental railroads, and again St. Louis became a nexus for the new mode of transportation.

The city was an early center of automobile development and manufacturing. In fact, there were nearly 100 automobile companies based in St. Louis in the early 20th century. Unsurprisingly, the nation's first gas station was opened in St. Louis in 1905. With this history behind it, it's befitting that the city has become known as the "gateway to the west." And it makes symbolic sense that Route 66 would pass right through it.

The St. Louis Art Museum contains a world-class collection of artworks and artifacts.

(Top left) **Busch Stadium opened in downtown St. Louis in 2006. It is the home of Major League Baseball's St. Louis Cardinals.**

(Top right) **The Anheuser-Busch St. Louis Brewery is now a national landmark. The brewery was first opened in 1852.**

(Left) **Until 2020, the Eat-Rite Diner stood on a prominent corner of the route, serving up diner classics to hungry travelers for decades.**

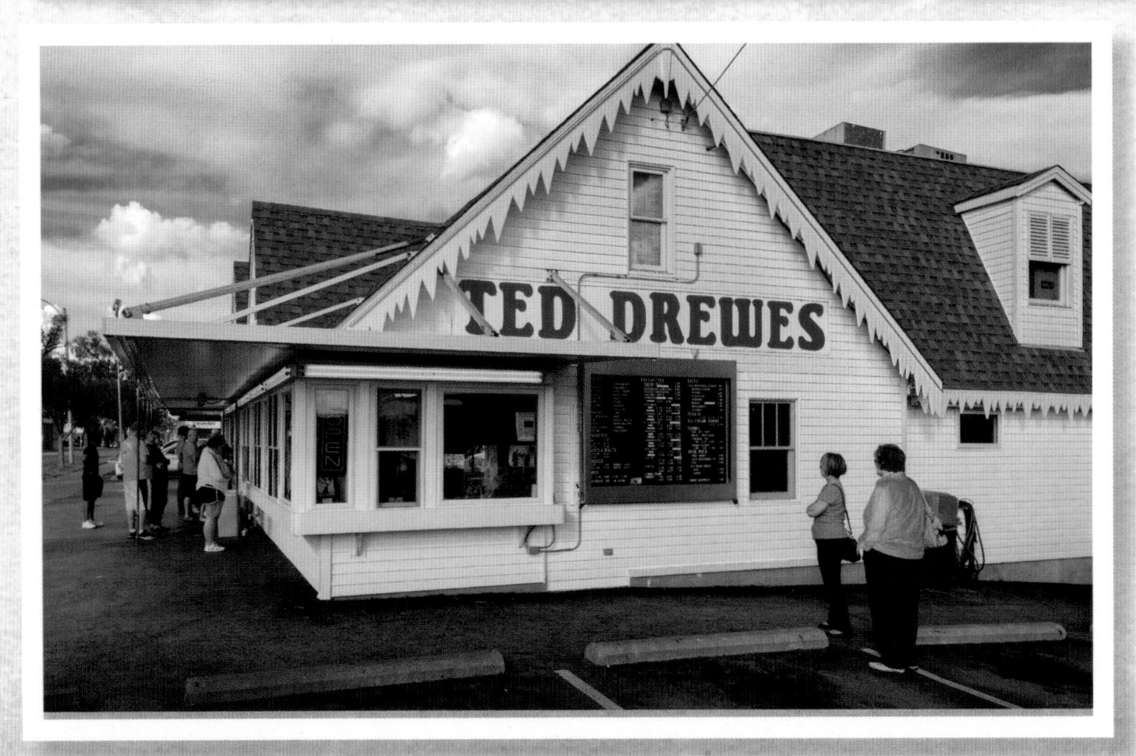

(Top left) **Stop off in Stanton to visit a museum dedicated to all things Jesse James.**

(Top right) **The Missouri History Museum in St. Louis is run by the venerable Missouri Historical Society, founded in 1866.**

(Left) **Before leaving St. Louis, be sure to stop for a little sugar for the road. Ted Drewes Frozen Custard has been selling chilly treats for over 80 years.**

The Missouri Botanical Garden in St. Louis was founded in 1859 and is the nation's oldest botanical garden.

The Meramec Caverns are located a few miles south of Stanton. These limestone caves enjoy a high profile along Route 66 thanks to the many signs advertising them. Guided tours include a computerized LED light show.

Slow down and enjoy the murals as you pass through Cuba.

In Cuba, Bob's Gasoline Alley once held hundreds of pieces of vintage gas station memorabilia. It closed in 2020.

(Top left) Cuba's FourWay has been both a filling station and a cozy diner.

(Top right) This Fanning attraction held the world record for largest rocking chair until 2015.

(Left) The Fanning 66 Outpost stands near the famous giant rocker.

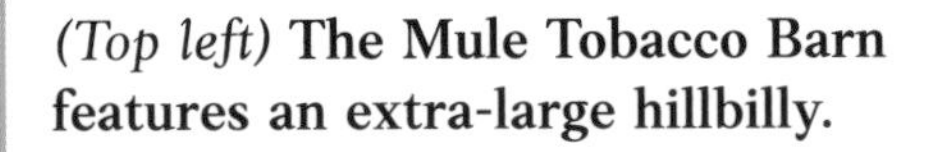

(Top left) **The Mule Tobacco Barn features an extra-large hillbilly.**

(Top and bottom right) **Devil's Elbow Bridge crosses the Big Piney River into the tiny village of Devil's Elbow. The bridge and village get their name from a particularly devilish bend in the river.**

Route 66 continues to change over time as sections deteriorate. This bridge on the Gasconade River, less than 30 miles down the road from Devil's Elbow Bridge, was finally closed and replaced in 2014.

Two classic motel signs in Lebanon. Both of the motels are still in business.

Springfield is the largest city in southern Missouri and considered the birthplace of Route 66.

(Top left) The Route 66 Springfield Visitor Center is filled with maps, guides, mementos, and all things Route 66. There's also a replica of a 1950s diner inside.

(Top right) There are plenty of places to stop and stretch your legs while in Springfield, including Jordan Valley Park.

(Left) One of the few remaining drive-in theaters in Missouri, the 66 Drive-In is right on the route, just west of Carthage.

(Left) **This tiny gas station was built in the 1920s and was originally located in the town of Avilla, east of Carthage.**

(Bottom) **About 20 miles west of Springfield, Paris Springs Junction maintains this replica of a 1930s gas station which once stood nearby.**

Kansas and Illinois are the only two states that can boast already having their entire lengths hard-surfaced when the route was first commissioned in 1926.

Kansas

Kansas contains just a sliver of Route 66 mileage—it has never exceeded 13 miles—but has definitely kept the historical spirit of the route alive. It runs west from the town of Galena in the southeastern corner of the state and then turns due south, passing through Baxter Springs. But packed into this tiny segment of the route is a series of great historical attractions.

On the outskirts of Galena is a restored Kan-O-Tex service station known as Cars on the Route. In addition to selling memorabilia, sandwiches, and snacks, Cars on the Route features some Pixar-themed vehicles on the surrounding grounds.

West of Riverton, the route turns south at the Rainbow Bridge. This little bridge is just wide enough for one-way traffic.

The future looks good for the Kansas stretch of Route 66: local support is strong and preservation efforts are ongoing.

CHAPTER 4

Oklahoma is considered one of the best states for driving Route 66. It contains more still-drivable route miles than any other state. And a drive through Oklahoma provides a deeply satisfying backdrop of pure Americana. The state is filled with charming small towns, quirky roadside attractions, refurbished gas stations, and classic diners.

Somewhat like the endless small farm towns of the Illinois segment, Oklahoma promises a laid-back litany of names—Chelsea, Kellyville, Chandler, Bridgeport, Elk City, Sayre—strung along outside the big towns of Tulsa and Oklahoma City. The route follows a diagonal path through rolling hills and plains into the center of the state. After Oklahoma City, the route runs west, crossing into Texas via the tiny town of Texola.

The Oklahoma segment also has bittersweet associations. John Steinbeck nicknamed the route "the road of flight" in *The Grapes of Wrath*. As the hard times of the Depression and the Dust Bowl overtook Oklahoma and surrounding states, Route 66 carried hordes of impoverished farmers and workers westward. Here in the middle, people remember that Route 66 was also a means of escape for the dispossessed.

Oklahoma is cattle country. Pastures of grazing cattle become a common sight driving through the state. As of 2020, the total number of cows was estimated to be over 5 million.

Route on the Range

When Route 66 was created in 1926, Oklahoma had little in the way of transportation infrastructure. There were still more railroad tracks than roads. Only about 10% of the state's roads were paved. Route 66 represented the change to come.

Allen's Fillin' Station in Commerce was built circa 1930. Now refurbished, it operates as a tiny gift store.

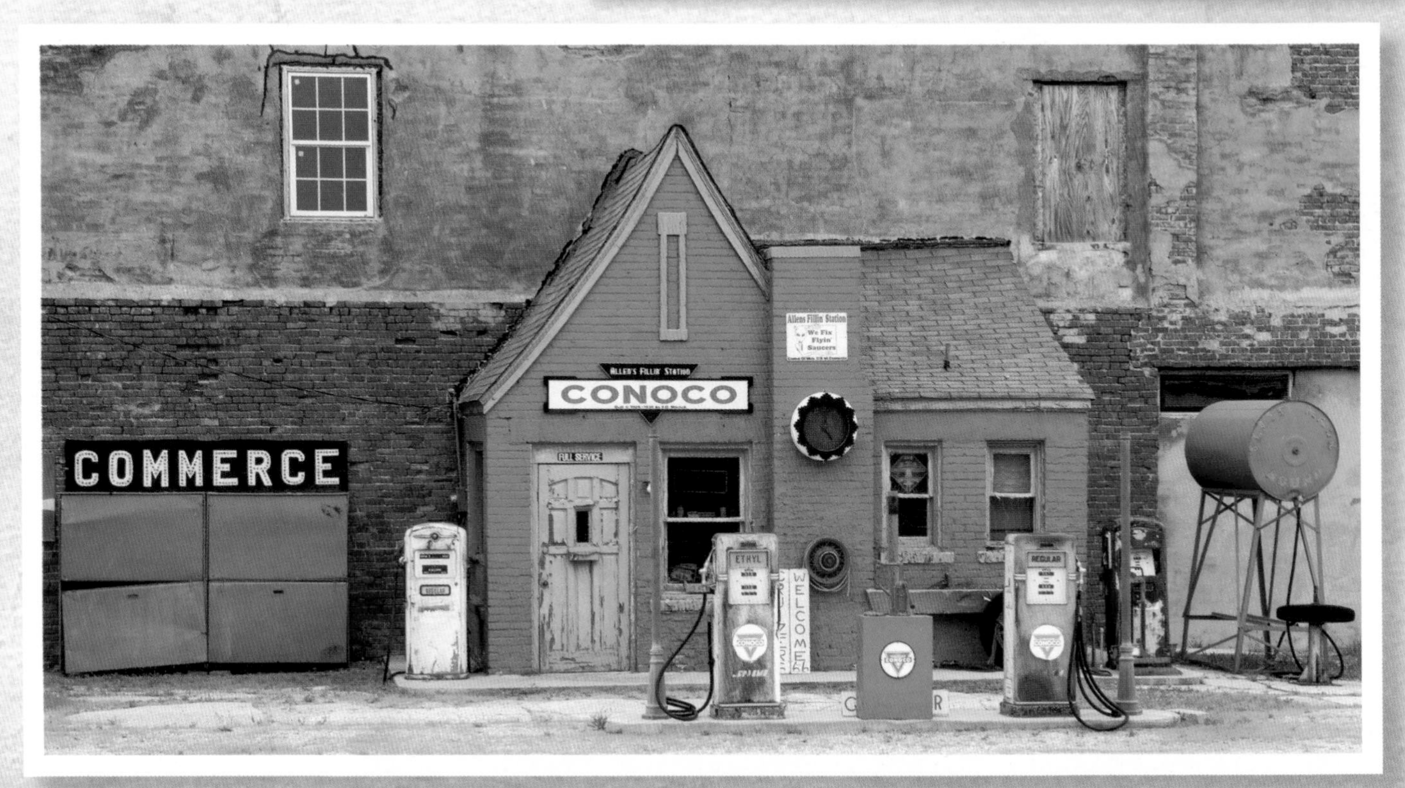

Miami is home to nine Native American tribal headquarters. Miami's Dobson Museum contains collections of artifacts from a number of local tribes.

(Right) Ed Galloway's Totem Pole Park is about three miles east of Route 66 where it passes through Foyil. The largest totem pole in the park, seen here, is labelled the "World's Largest Concrete Totem Pole." It took Ed Galloway 11 years, adding one bucket of cement at a time, to create his masterpiece. It has more than 200 different carvings on its facade, and its base is a stout 18 feet in diameter.

(Bottom) Downtown Chelsea has a quiet Main Street feel. The town boasts the location of the first oil well in the state, though the claim is disputed.

Catoosa's famous blue whale was originally built by Hugh Davis as an anniversary present for his wife in the early 1970s. The cement whale attracted so much attention on the route that it became the showpiece for a reptile zoo and swimming hole operated by Davis until shortly before his death. The community has since rallied around the landmark, refurbishing it and applying thousands of gallons of blue paint in the process.

About 100 miles into Oklahoma, Route 66 arrives in Tulsa. The bustling city is situated on the Arkansas River and is home to about 400,000 people.

The Cyrus Avery Centennial Plaza stands along the route in downtown Tulsa. Avery is often called the "Father of Route 66" because of his role in the route's creation. He helped plan a national system of numbered highways and proposed a highway linking Chicago and Los Angeles. Avery founded the U.S. 66 Highway Association.

Tulsa Art Deco

With a nickname like "Oil Capital of the World," it makes sense that Tulsa would be part of the Oklahoma segment of Route 66. Oklahoma's oil boom began in the early 20th century, and as the industry flourished, so did Tulsa. Along with the giant storage tanks, derricks, and pipelines, the city put up fine Art Deco buildings, showing off its new wealth. The Boston Avenue United Methodist Church, Oklahoma Natural Gas Building, and Tulsa Union Depot, are all examples of Tulsa Art Deco.

(Top) **Closeup of the exterior of the Oklahoma Natural Gas Building.**

(Left) **Tulsa Union Depot is now home to the Oklahoma Jazz Hall of Fame.**

The Boston Avenue United Methodist Church.

(Top left and right) **Tulsa's** ***Golden Driller*** **is a 75-foot-tall statue commemorating petroleum industry workers.**

(Left) **Tulsa's Woody Guthrie Center commemorates the great American folksinger.**

(Right) **Old cars commemorate the route through downtown Bristow.**

(Bottom) **Another classic main street in Stroud.**

Retro signs, buildings, and shops in Stroud.

Jimmy's
ROUTE
66
ROUND-UP CAFE
Round-up
Café
Chicken Fried Steak,
BBQ Ribs, Catfish
ROUTE
66
SINCLAIR
PENNSYLVANIA
MOTOR OIL
ENJOY
RC
Royal Crown Cola
ICE COLD
SOLD HERE
Coca-Cola
RESERVED
PARKING

(Left) **A tiny renovated Phillips 66 service station along the route in Chandler.**

(Bottom right) **A small souvenir shop located right next to the Arcadia Round Barn. The building also houses a neon-lighting company.**

(Bottom left) **Arcadia, just north of Oklahoma City, is home to the famous Arcadia Round Barn. The barn was built in 1898 and fully restored in 1992. It is still one of the most photographed buildings on the route.**

Arcadia's Pops is a restaurant that can't be missed from the road. The 66-foot illuminated soda bottle sculpture beckons parched travelers to stop in and try one of the hundreds of varieties of soda available. The location serves as a diner, gas station, and convenience store.

Oklahoma City

Oklahoma's state capital has plenty of route-related attractions, historic buildings, memorials, motels, and vintage gas stations. With its strong western American culture, it's not surprising that the city is also home to the National Cowboy & Western Heritage Museum, the American Banjo Museum, the Oklahoma Railway Museum, the American Indian Cultural Center, and even a rattlesnake museum.

(Top left) Before there was the route there was the trail: Frederic Remington's statue of exuberant cowboys stands outside the National Cowboy & Western Heritage Museum.

(Top right) The Oklahoma State Capitol Building.

(Right) Stop off at the Milk Bottle Grocery while you're in town. The giant milk bottle has been resting atop the building since 1948.

The National Cowboy & Western Heritage Museum was founded in 1955. The museum exhibits western art and artifacts. Its exhibition wing features a turn-of-the-century cattle town, complete with creaking wooden walkways and ambient sounds. Interactive galleries focus on all things western: rodeos, firearms, Native American culture, and of course, the American cowboy. The art gallery features works from the likes of Frederic Remington, Charles Russell, and William Robinson Leigh. The pictured statue of two mounted cowboys shaking hands is titled, “Code of the West.”

(Right) **The William H. Murray Bridge (also known as the Pony Bridge) is one of the most notable bridges along the route and is nearly 4,000 feet long. The bridge marks a change in scenery as the route enters the western plains.**

(Bottom) **So long, OKC. The Lake Overholser Bridge served as Route 66's bridge leaving the city for over three decades. The bridge still serves local traffic.**

The Oklahoman sharecroppers in John Steinbeck's *The Grapes of Wrath* lose their home and must pack up their belongings and journey along Route 66 to California. This fictional story reflected the hard times that millions experienced during the Great Depression and the Dust Bowl. Steinbeck found a potent American symbol in "the migrant road," because it spoke to the experiences of so many. Route 66 became the road on which Americans from diverse backgrounds could unify. It was a path, a home, and an identity.

During the Depression years, the route also provided employment opportunities to Americans across the country. President Roosevelt's New Deal programs funneled money into road improvements and maintenance work. From 1933 to 1938, thousands of people worked on road gangs. Because of this, the entire highway from Chicago to Los Angeles was paved by 1938.

(Top left) Lucille's was a little gas station and motel outside of Hydro. It served customers from 1929 until 2000.

(Top right) A marker at Lucille's commemorates the Will Rogers Highway, another of the route's nicknames. Native Oklahoman Will Rogers played a major role in popularizing the route through his syndicated newspaper columns.

(Left) Canute's Cotton Boll Motel has closed, but the aging sign remains in nostalgic splendor.

(Top and bottom right) **Elk City's National Route 66 and Transportation Museum gives visitors a chance to share the experiences of those who traveled and lived along the route via recorded histories and personal accounts.**

(Bottom left) **The Old Town Museum is right next door. This museum focuses on the life of early Oklahoma pioneers and features period buildings like the Pioneer Chapel.**

(Top) **A whimsical mural decorates the side of a local watering hole in Sayre.**

(Left) **The Sandhills Curiosity Shop in Erick is indeed packed with curiosities. Nothing is for sale here, but it's worth stopping just for the ambience the owners have created.**

Antique cars rust in a field near Texola. Nearly a ghost town with only handful of people still remaining, Texola is the last town before the Texas state line.

Texas
STATE LINE

CHAPTER
5
Texas

Route 66 draws a westerly line across the Texas Panhandle, entering near Benonine and crossing into New Mexico at Glenrio, a ghost town that straddles the border. The Texas Panhandle is the southern end of the Great Plains. Route 66 stretches nearly 200 miles across this mostly flat and dry landscape dotted with grain silos and cattle ranches. Giant windmills increasingly cover the region—wind power is a major source of electricity in the state.

Small towns sparsely pepper the route. Amarillo is the only real city encountered on the Texas segment. Unlike previous states, roadside attractions here can be few and far between. Instead, prepare to experience a vast and hypnotic prairie where you might still spot a buffalo.

(Top left) **The first stop in Texas is Shamrock, home to the Tower Station and U-Drop Inn, a historic café that first opened in 1936.**

(Top right) **An aging mural in McLean. The town made the most of Route 66's golden age and offered passing motorists a choice of gas stations, motels, cafés, and restaurants.**

(Left) **Built in 1928 and renovated in 1992, this little "Cottage Fashion" service station was the first Phillips gas station to operate in Texas.**

(*Left*) A derelict service station in the near-ghost town of Alanreed.

(*Bottom right*) Known as The Leaning Tower of Britten, this water tower tilts precariously outside Groom. It was erected, complete with 10-degree tilt, to attract motorists to the local truck stop nearby.

(*Bottom left*) The Slug Bug Ranch in Conway features brightly-painted cars partially buried in the ground. Visitors have been coming here for years to add finishing touches with spray paint, blowtorches, and crowbars.

Cool Your Heels in Amarillo

The only major city on the Texas segment, Amarillo quickly developed a reputation as an oasis point on the route. Gas stations, motels, and the "Amarillo City Tourist Camp" sprang up as route traffic blossomed. Amarillo began as a cattle town, but it rode the prosperity boom of Route 66, and managed to avoid the subsequent bust as I-40 changed the game for local economies.

Amarillo's U.S. Route 66-Sixth Street Historic District is worth making a stop for. Along with its collection of commercial buildings that hold historic significance for the route, the district also has art galleries, antique stores, and specialty shops.

The Big Texan Steak Ranch has been serving up enormous steaks and assorted Texan grub since 1960.

(Top left) **The RV Museum at Jack Sisemore Traveland is a collection of unusual RVs.**

(Top right and bottom left) **One of the most iconic stops along the Texas segment has to be the Cadillac Ranch just west of Amarillo. A row of cars with their front ends buried in the ground stands covered in layer upon layer of spray paint (the public is free to add more layers). The installation was created in 1974.**

While in the Neighborhood

Palo Duro Canyon State Park is less than 30 miles down the road from Amarillo and well worth the detour. It contains the second-largest canyon in the country. This beautiful area is readily accessible by car and offers some of the most spectacular landscapes in the Texas Panhandle.

The canyon system of Palo Duro is about 120 miles long. The canyon is famous for its caves and hoodoos. Activities in the area include hiking and biking trails and a zip line.

(Top left) An intersection in Adrian. The endless plains keep rolling on.

(Top and bottom right) Those traveling the entire route can get a sense of perspective in Adrian. Another tiny Panhandle town, Adrian is home to just over 160 people. The midpoint of the route is well served by the MidPoint Café.

The last Texas town on the route stands on the border of New Mexico. Glenrio is now a ghost town consisting of a few derelict buildings.

CHAPTER 6

New Mexico

New Mexico presents the traveler with a series of changing landscapes. The vast ranchlands of the plains begin to disappear, and in their place come mesas, pine forests, sudden hills, and beautiful mountain ranges. The view from the window is worth the drive alone.

Route 66 had a rough start in New Mexico. In 1926, the state was barely 14 years old. The route had to be patched together from highway segments, gravel roads, and former wagon trails. The result was a meandering line that topped 500 miles. Over the following decades, realignments reduced the mileage and straightened the route. Federal highway building in the 1950s led to I-40, which made use of much of Route 66 in New Mexico.

Today there are still over 260 miles of pre-interstate Route 66 that are drivable, though in this state it's not always easy to recognize them. Travelers can help themselves out by picking routes ahead of time (such as the state's Route 66 National Scenic Byway) and arming themselves with a few good maps and road guides.

(Top right) The New Mexico segment begins with more plains than peaks.

(Right) The first time change of the route happens at the New Mexico state line.

(*Top left*) Tucumcari's Tee Pee Curios is an old gas station converted into a souvenir shop.

(*Top right*) An old section of the route near Newkirk.

(*Left*) An old Esso service station in Tucumcari.

The historic Blue Swallow Motel in Tucumcari is one of the longest continuously operating motels on the New Mexico segment. It was built in 1941.

Santa Rosa's Guadalupe County courthouse was built in 1909.

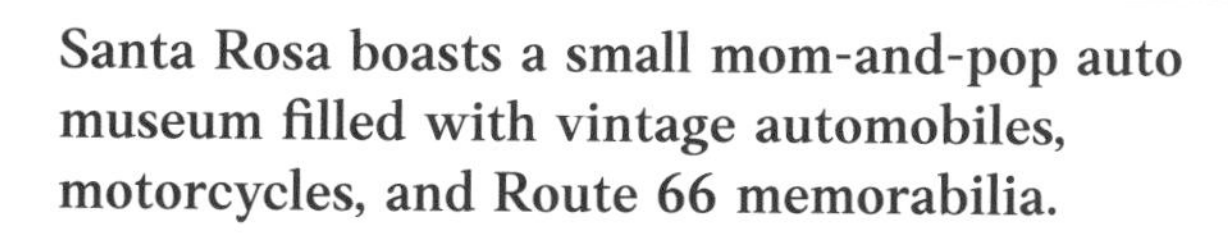

Santa Rosa boasts a small mom-and-pop auto museum filled with vintage automobiles, motorcycles, and Route 66 memorabilia.

The Old Santa Fe Loop

The first New Mexico Route 66 alignment followed a snaking S curve across the center of the state. Just west of Santa Rosa, it turned north, taking a convoluted course to Santa Fe before turning southwest toward Albuquerque. It passed south through Albuquerque to Los Lunas before turning west again. This meandering was engineered out of existence in subsequent decades, but though efficiency was gained, some of the charm was lost.

Santa Fe is like no other city on Route 66. It was established in 1610 by Spanish colonists, became part of Mexico in 1821, and was ceded to the United States in 1848. The city preserves its own unique architectural style, art, and cuisine.

The Santa Fe Loop is definitely worth the drive. It first runs north across arid plains, eventually entering hill country which provides enchanting views at every turn. It follows the Pecos River Valley to Santa Fe—a city definitely worth stopping for. The road from Santa Fe to Albuquerque shows how much the landscape has changed since the plains of Oklahoma and Texas.

(Left) Pecos National Historical Park, on the route to Santa Fe.

(Bottom right) Santa Fe's Cathedral Basilica of Saint Francis of Assisi.

(Bottom left) In Santa Fe (and northern New Mexico), strings of local chiles are hung over doors, portals, fences, and patios. Called *ristras*, these colorful strings are hung up to dry for later cooking, but are sometimes used simply for decoration.

Downtown Santa Fe offers plenty of opportunities to find a uniquely regional souvenir.

Albuquerque

Albuquerque is roughly in the center of New Mexico. It sits along the Rio Grande River in a broad valley partially enclosed by the Sandia and Manzano mountains. The city holds about half a million people.

The first route alignment of 1926–1937 ran through the city from north to south. After realignment, it entered the city from the east, running along Central Avenue. Driving along Central, travelers experience an 18-mile urban stretch that boasts plenty of decades-old motels and motor courts, diners, city attractions—and city traffic.

Motel signs in Albuquerque.

(Top left) **Albuquerque's 66 Diner venerates the route's bygone days with classic milkshakes, blue plate specials, retro décor, and an oldies jukebox.**

(Top and bottom right) **The Albuquerque International Balloon Fiesta takes place every October.**

Rugged mesa landscape between Mesita and Laguna. As Route 66 enters the last third of the New Mexico segment, mesas and rugged hills become more common.

(Top left) **The historic village of Laguna Pueblo is about 45 miles west of Albuquerque.**

(Top right) **A wind-worn sign outside San Fidel.**

(Right) **A view from the road outside Thoreau.**

A famous formation just north of the route in Church Rock.

The only sizable town between Albuquerque and Flagstaff, Gallup was founded in 1881 when the Santa Fe Railroad built a line in the area. From its earliest years, it was known as the Gateway to Indian Country.

Colorful mesas and plains mark the final segment of the route in New Mexico.

(Top) Crumbling structures dot the barren desert landscape. Although the weather is typically hot and dry, latitude affects temperature. Nighttime temperatures can plummet. Travelers staying in the area will often need warm clothes if they're out at night.

(Left) Travelers find themselves in a truly enchanting desert environment in western New Mexico. Weathered rock hoodoos stand eerily above a landscape of arid canyons and plains.

(Left) The original 1926 alignment of Route 66 crosses the Continental Divide. Though the spot is somewhat forgotten now, in former years it boasted a number of attractions, including the Continental Trading Post, the Great Divide Trading Company, and the Top O' the World Hotel and Café.

(Bottom) Gallup lacks the scenic glory of former days but is a fascinating place, nonetheless. It was once a railroad nexus and is still the main commercial center for the surrounding Navajo homelands. Rodeos and powwows continue to be held here, and the annual Inter-Tribal Ceremonial is held in nearby Red Rock Park every August. After dark, it's worth driving along its nostalgia-inducing neon-lit motel row.

CHAPTER 7

Arizona

Arizona provides one of the most dramatic stretches of Route 66. Deserts, mesas, volcanoes, and even lush forests are dotted with ghost towns, trading posts, and mysterious ruins. Painted deserts, the Petrified Forest, Meteor Crater, and the Grand Canyon are also near the route. Towns can be few and far between. Route 66 enters some of its highest elevations here; just west of Flagstaff the route reaches a high point of over 7,300 feet at the Arizona Divide.

About 400 miles of Route 66 passed through Arizona in 1926, and in the beginning, practically none of it was paved. The route was "completed" by 1938, but as elsewhere, Arizona's total mileage was gradually whittled down as dangerous curves were straightened out and busy towns were bypassed. Even though I-40 has replaced much of the route, Arizona can still boast long continuous stretches of uninterrupted route—particularly the sections between Ash Fork and the Colorado River.

Lupton is the first town inside the Arizona border. Soaring sandstone cliffs stand above the town. Travelers stopping for a picture of the Painted Cliffs will also find souvenirs aplenty.

(Top) **The Querino Canyon Bridge was built in 1929 along an older alignment of the route. It still carries local traffic on the Navajo Indian Reservation.**

(Right) **The Petrified Forest National Park is about 50 miles from the New Mexico border. It's a must-stop along the route. The park gets its name from the large deposits of petrified wood to be found across its desert landscape. The park was established as a national monument in 1906.**

The amazingly lifelike logs abundantly found in the park are the crystallized remains of trees that grew around 225 million years ago. These trees were buried by sediment and protected from decay. Over the eons, the original plant material was replaced as mineral-rich groundwater seeped into the wood cells. The logs now consist of nearly solid quartz.

Holbrook is home to the famous Wigwam Motel.

A shop in Holbrook attracts visitors with its immense concrete dinosaur.

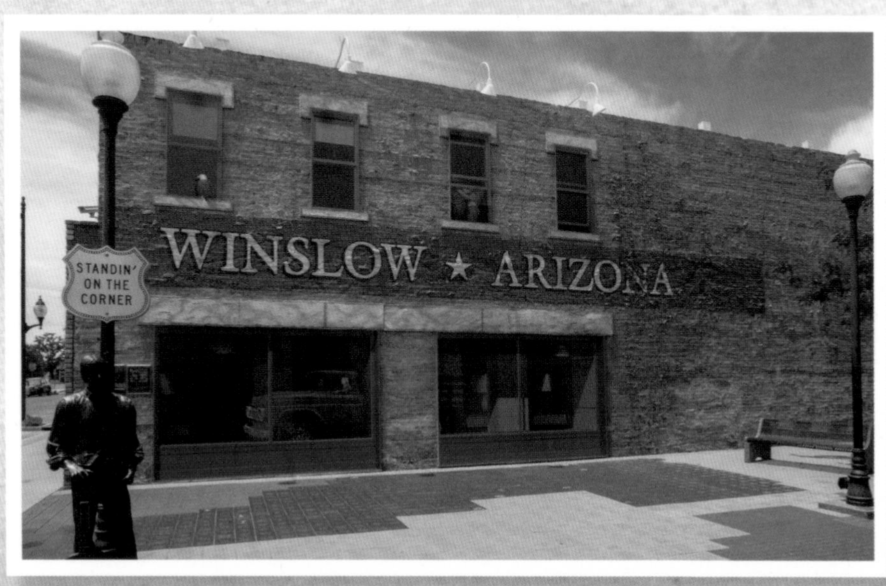

Downtown Winslow's dedication to the famous Eagles song, "Take It Easy" is a must-see.

Meteor Crater is about 18 miles west of Winslow and 6 miles south of the route.

(*Left*) The Twin Arrows Trading Post no longer exists, but the arrows remain.

(*Bottom*) Weathering ruins along the route west of Winslow.

Flagstaff

The Flagstaff area is worth slowing down for. There are plenty of outdoor activities: Coconino National Forest, Elder Pueblo Archaeological Site, Red Rocks State Park, and Walnut Canyon National Monument are all in the vicinity. The town itself celebrates Route 66 with plenty of murals, diners, and neon.

(Top left) **The historic train depot was built in 1926.**

(Top right) **Flagstaff's Hotel Monte Vista, just off Route 66, first opened in 1927.**

(Left) **View from the route: classic southwestern landscape outside Flagstaff.**

(Left) **The little mountain town of Williams is known as a gateway to Grand Canyon National Park.**

(Bottom right) **Pete's Gas Station Museum in downtown Williams.**

(Bottom left) **Route 66 is still Main Street in Williams, and it packs plenty of nostalgia into a few town blocks.**

The Mother of All Side Trips

The Grand Canyon National Park is about an hour's drive from Williams. It's a side trip too magnificent to pass up. Drivers ready for a respite can even take the daily Grand Canyon Railway from Williams directly to the South Rim.

"The wonders of the Grand Canyon cannot be adequately represented in symbols of speech, nor by speech itself," said Major John Wesley Powell, the first to explore the Colorado River through the canyon. And more than a century ago, the sight stunned Theodore Roosevelt into silence.

Although there is no consensus on the Seven Natural Wonders of the World, the Grand Canyon is on virtually every short list—sometimes the only such landmark in the United States. The vast majority of tourists who visit the park see the canyon from the South Rim's Grand Canyon Village Historic District. There the canyon reaches ten miles across and one mile down. Visitors hiking from the rim all the way down to the river will traverse seven miles. In other parts of the park, however, the Grand Canyon descends another 800 feet farther and stretches eight miles across.

More than four million visitors come to see the canyon every year, seeking its beauty beyond compare. That's more than visit the Lincoln Memorial, even though the Grand Canyon is much farther from most U.S. population centers. Visitors explore the Grand Canyon by boat and raft, horse and mule, car and bus, or helicopter and airplane.

The 1,450-mile-long Colorado River, the wellspring of the canyon's origin, flows west at about four miles per hour. The canyon, of course, is the jewel of the Colorado River, sitting on the borders of the Havasupai, Navajo, and Hualapai reservations. A hike from the rim to the riverbed is an overnight hike from any entrance.

A lonely stretch of the old route outside Ash Fork.

(Left) The desert has partially reclaimed a former gas station near Ash Fork. The dry desert climate eradicates human presence more slowly than other environments. Because of this, the Arizona segment of the route is dotted with enigmatic ruins.

(Bottom) An abandoned cattle corral withers away outside Ash Fork.

Tiny Seligman (population: 456) is the beginning of an uninterrupted stretch of historic Route 66.

(Top and bottom right) **The Hackberry General Store stands in isolated splendor on the historic route between Seligman and Kingman.**

(Bottom left) **West of Seligman on the historic route.**

Kingman greets travelers with classic motels, service stations, diners, and vintage kitsch.

Route 66 winds through a gap in the Black Mountains known as Sitgreaves Pass. This section has hairpin turns and awesome views around every bend.

After the excitement of the Sitgreaves Pass, Oatman is a great place to stop, stretch, mail a postcard, and check out some souvenirs. Oatman was once a classic western gold mining town.

(*Left*) The route passes near Goose Lake, just outside of Topock, as it leaves Arizona.

(*Bottom*) South of Topock, on the California border, Lake Havasu appears in stunning contrast to its desert surroundings. The lake is a large reservoir formed by the construction of the Parker Dam in 1938. The Colorado River flows into the lake, making a desert oasis of palm trees, sandy beaches, and craggy mountains.

The end of the Arizona segment comes with spectacular views of river, lakes, and marshes nestled among the mountains.

ROUTE
66
CHAPTER
8
California

Route 66 crosses into California and turns north along the Colorado River, passing through Needles before making a westward turn into the Mohave Desert. This may be the most daunting section of the entire route. Sandy plains, mesquite brush, and barren mountains shimmer in the heat of the driest desert in the country. This harsh landscape has its attractions though—abandoned mining settlements, ghost towns, and unusual vegetation provide vistas encountered in no other state.

California is fortunate to have most of its original segment intact. Signs mark most of its 300-plus miles. Travelers experience every kind of southern California landscape on this last segment; the route runs through mountains, desert, lush valleys, and a long stretch of urban Los Angeles on the final stretch to the Pacific Ocean.

(Top right) **The Joshua tree is an iconic presence in the Mohave Desert. This unusual-looking plant is not actually a tree—it's a succulent belonging to the yucca genus.**

(Right) **Route 66 used the Old Trails Bridge to cross the Colorado River into California until 1948. The bridge was first opened to traffic in 1916.**

Needles has plenty of vintage motels and things to see. It's a good place to stop before crossing the Mojave stretch. But be warned: summertime temperatures here stay above 100 degrees for weeks at a time.

Mojave National Preserve

Route 66 passes to the south of this immense preserve of desert wilderness. 1.6 million acres in size, the area is home to mesas, mountains, canyons, sand dunes, cinder cone volcanoes, and the fading relics of brief human settlement. Joshua tree forests, creosote bushes, cholla cacti, pinyon pines, yucca, and juniper are just a few of the unique and hardy plants found in this surprisingly diverse environment.

The golden dunes, rugged rocks, and solitary splendor of the Mojave Desert give this part of Route 66 a mystique found nowhere else. There are plenty of long, lonely stretches filled with nothing but the dusty road, distant mountains, and desert sky. It's the driest place in North America, and possibly the quietest too.

The Mojave National Preserve's Kelso Dunes lie north of the route, but well worth a detour. A vigorous hike into the dunes reveals a surreal backdrop unlike anyplace else. Here, you can hear the eerie sounds of "singing sands," a natural sound also described as squeaking, roaring, whistling, or booming. The dunes cover about 45 square miles and rise to 650 feet above the desert floor.

(Left) **The desert's beautiful cholla cactus, with its dense covering of barbed spines, is best admired from a distance.**

(Bottom right) **High desert. Even if only seen through a car windshield, the desolate beauty of this area is unforgettable.**

(Bottom left) **Rugged, long-lived creosote bushes dominate the ecosystem of the Mojave. These plants can survive a season or two of no water at all and some plants may live for thousands of years.**

Roy's in Amboy was first opened in 1938. In its heyday, Roy's featured a gas station, auto repair shop, café, and motel.

(Top) **A desert vista in the vicinity of Ludlow.**

(Left) **An iron dinosaur rusts near Ludlow.**

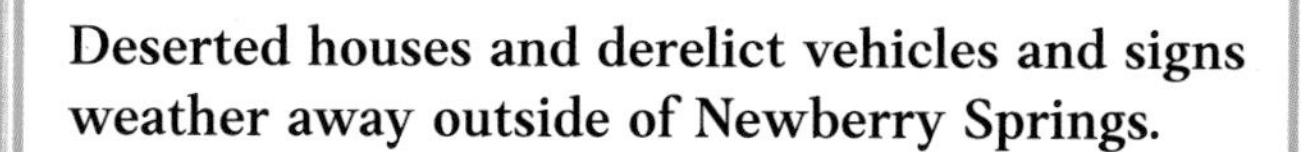
Deserted houses and derelict vehicles and signs weather away outside of Newberry Springs.

(Top left) Just before reaching Barstow, turn off the route to visit the Calico Ghost Town. Once a silver mining town, the area now offers a glimpse of former boom years. Several buildings have been reconstructed and visitors can ride a short segment of the railroad.

(Top right) Antique cars sit in front of the Route 66 Motel in Barstow.

(Left) Barstow features plenty of motels, restaurants, and a Route 66 "Mother Road" Museum.

(Top left) After the endless desert miles, the bustling town of Barstow allows drivers a brief taste of civilization and an opportunity to refresh and refuel. Barstow was a mining town in the 19th century and later developed into a transportation hub. It's now an important midway stop between Las Vegas and Los Angeles.

(Top right) Route 66 motels are alive and well in Barstow. It's a good place to stop for the night—there is still plenty of desert road awaiting drivers.

(Left) Mile after mile of isolated, desert splendor.

(Top and bottom left) **The colorful Elmer's Bottle Tree Ranch appears like a desert mirage outside Helendale.**

(Bottom) **The desert stretch ends in dramatic fashion as the route approaches Cajon Pass. Gaining elevation, it passes between the San Bernardino and San Gabriel Mountains.**

(Top left) After Cajon Pass, the route takes on a new look. It descends towards San Bernardino and the sprawl of the Los Angeles basin. Route 66 runs south from the pass until it turns west again in San Bernardino. From here, the route is headed to the ocean!

(Top right) Route 66 passes through Rancho Cucamonga, offering a spectacular view of the nearby San Gabriel Mountains.

(Right) In San Bernardino, traffic congestion is to be expected, but so are southern California sunsets.

Los Angeles

Route 66 followed quite a few different alignments through the city over the years. A convenient way to follow it now is to follow the prominent road signs. Head west from San Bernardino along Foothill Boulevard into Pasadena. Follow the signs south through downtown Los Angeles, and then Hollywood, where the route joins Santa Monica Boulevard. The road ends at Ocean Avenue, a few blocks north of the Santa Monica pier. The pier itself has a brass plaque marking the official end of the route.

(Top) Running through Pasadena, Route 66 passes the Fair Oaks Pharmacy, a soda fountain and diner that has been a stopping point along the route from the very beginning.

(Left) Dodger Stadium is north of downtown Los Angeles.

Route 66 passes through roughly 80 miles of the metropolitan area around Los Angeles.

(Top left) The route passes the Hollywood Forever Cemetery, where many famous film luminaries are buried.

(Top right) As part of Santa Monica Boulevard, the route passes directly through upscale Beverly Hills.

(Left) Keep a sharp eye out while on Santa Monica Boulevard in Hollywood—on a clear day, the Hollywood sign can sometimes be seen through the trees and buildings to the right.

Though no part of Route 66 ever actually ended here, the Santa Monica Pier is now considered a classic end point for the journey.

1882
FLAGSTAFF
CONOCO
MOTEL
CAFE

ROUTE
US
66